BANG

To Charlie in 2007, meticulously measuring pebbles at Lulworth Cove.
To June in 1941, watching the northern lights from London during the Blackout.
To my brother and grandmother, and to all those who love learning. – J.S.

First published in the United Kingdom in 2024 by
Thames & Hudson Ltd, 181A High Holborn, London WC1V 7QX

Bang © 2024 Thames & Hudson Ltd, London
Text and illustrations © 2024 Jennifer N. R. Smith

Consultancy by Jon Cannell

British Library Cataloguing-in-Publication Data
A catalogue record for this book is available from
the British Library

ISBN 978-0-500-65334-0

Printed in China by Artron Art (Group) Co., Ltd

Be the first to know about our new releases,
exclusive content and author events by visiting
thamesandhudson.com
thamesandhudsonusa.com
thamesandhudson.com.au

JENNIFER N. R. SMITH

BANG

the WILD WONDERS of EARTH'S PHENOMENA

CONTENTS

A PHENOMENAL WORLD

BANG! A VOLCANO IN INDONESIA BLASTS HOT LAVA IN A SPECTACULAR FIERY SPRAY.
Meanwhile, thousands of miles away, far to the north, an arctic fox admires dancing
green rivers of light in the sky. The Earth is a phenomenal planet!

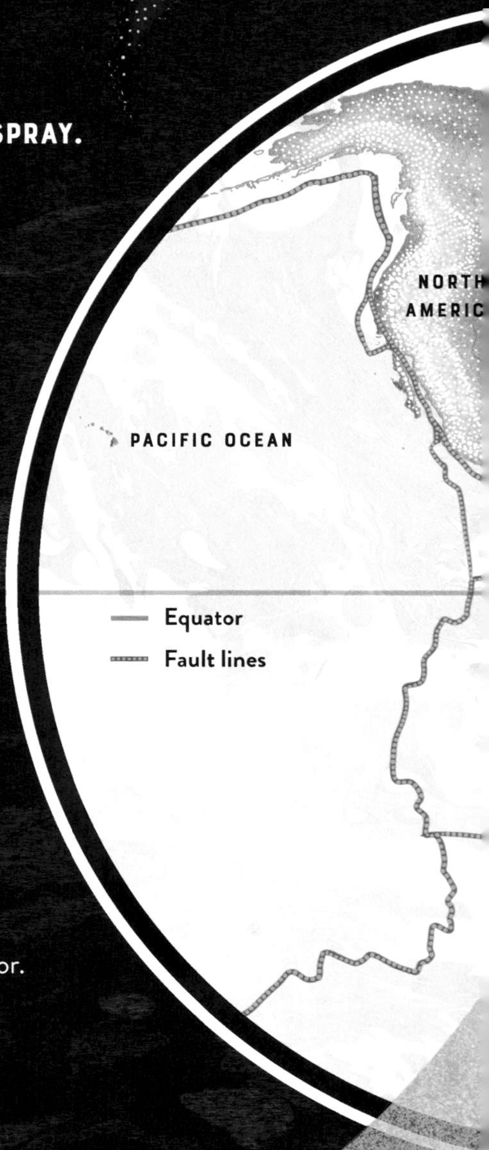

The word **PHENOMENON** has two meanings. The first relates to a feature or event that can be observed and experienced through the senses. It also means something that is remarkable or exceptional - something with great impact! The plural of phenomenon is phenomena.

Natural phenomena are found in nature, as opposed to being made by humans. Some phenomena, such as rainbows or lightning, are familiar sights in our daily lives. Others are rarer or only found in certain places, like volcanoes.

The types of phenomena you may experience often depend on the **climate** in your location and the underlying **geology** of the landscape.

Phenomena such as mountains, volcanoes, **geysers** and earthquakes are most common along tectonic plate lines (see pages 8–9). **Atmospheric phenomena** are more likely to occur in places that are either very cold – towards the North and South Poles – or very hot – towards the equator.

NORTH AMERICA

PACIFIC OCEAN

— Equator

----- Fault lines

PLATE KEY

1. ARCTIC FOX, NUNAVUT, CANADA 2. GEYSER, LAKE BOGORIA, KENYA
3. NARUTO WHIRLPOOLS, JAPAN 4. MOUNTAINS, ANDES, PERU
5. GLACIERS AND ICEBERGS, ANTARCTICA
6. VOLCANO, MOUNT MERAPI, INDONESIA

1.

2.

3.

ARCTIC OCEAN

EUROPE

ASIA

ATLANTIC OCEAN

AFRICA

SOUTH AMERICA

INDIAN OCEAN

AUSTRALIA
& OCEANIA

SOUTHERN OCEAN

ANTARCTICA

6.

5.

THE STRUCTURE OF EARTH

FANCY A SLICE? TO UNDERSTAND MANY OF THE PHENOMENA IN THIS BOOK, IT'S HELPFUL TO KNOW the structure of the Earth. If you were to cut yourself a slice, you would see that the Earth has a crust like a pie. But underneath this hard surface, it's not as solid as it seems! Similar to a pie fresh out of the oven, the further towards the centre of the Earth you go, the hotter it gets. So hot, in fact, that rock and metal melt and turn to liquid.

A SLICE OF THE EARTH

THE ATMOSPHERE

The outer layer of the Earth is called the atmosphere, which acts as a protective bubble. It is made of lots of different gases, including the oxygen we breathe. It is also where many impressive phenomena take place, such as storms and auroras!

THE CRUST

The Earth seems very solid because we live on the crust, which is made from hard rock. This is the thinnest layer of the Earth and it is split up into **tectonic plates**. These fit together a bit like a jigsaw puzzle.

THE MANTLE

The rock underneath the crust is called the mantle. It is hot and semi-solid, like the gooey filling in a pie. The mantle is in a constant cycle of heating, rising towards the crust, cooling and then sinking towards the core.

THE OUTER CORE

Underneath the mantle is the outer core, which is made of liquid metals. Both the outer and inner core are outrageously hot, around 5,200°C!

THE INNER CORE

At the very centre of the Earth sits the inner core – a ball of solid metals, nickel and iron. At an incredible 6,400 km under the surface, the metals are under so much pressure that they are unable to melt, despite the immense heat!

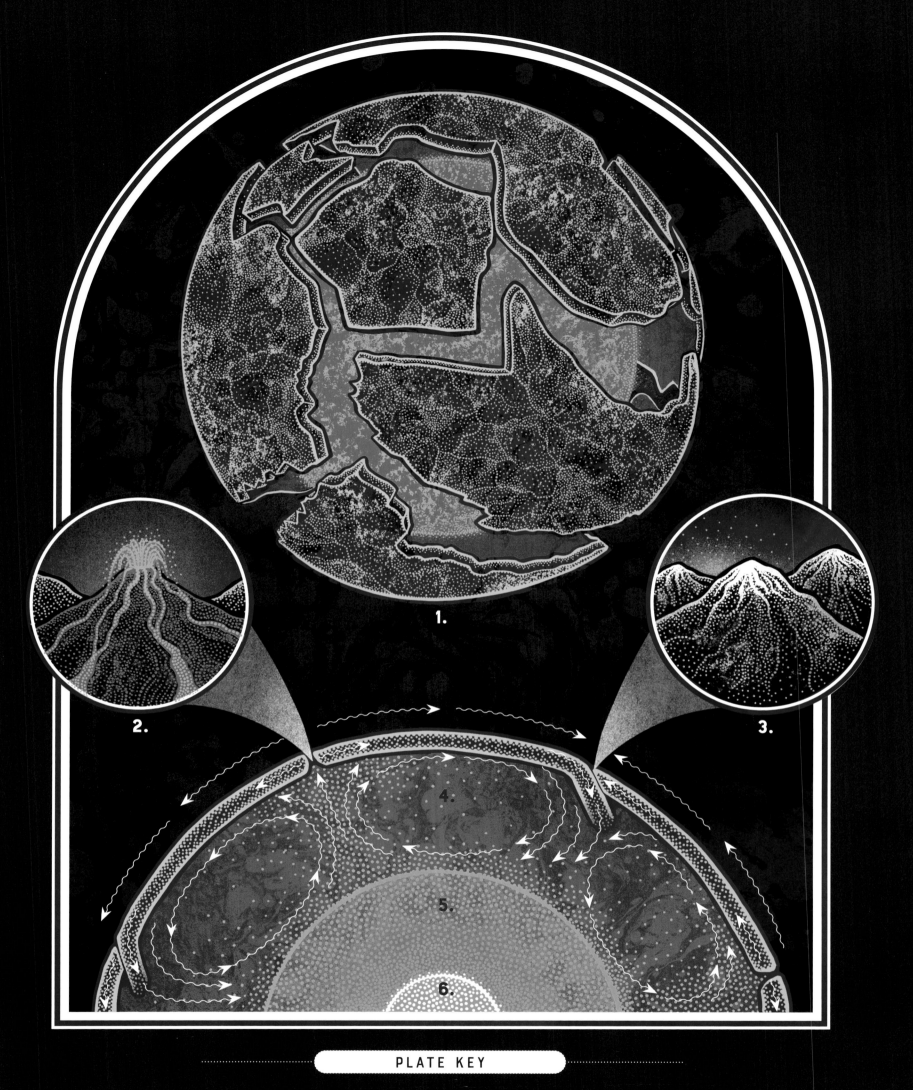

PLATE KEY

1. TECTONIC PLATES OF THE EARTH   **2.** VOLCANO FORMING FROM PLATES MOVING AWAY FROM EACH OTHER
3. MOUNTAIN RANGE FORMING WHERE PLATES MOVE TOWARDS EACH OTHER
4. LIQUID ROCK MOVING IN THE MANTLE   **5.** THE OUTER CORE   **6.** THE INNER CORE

The crust of the Earth is split up into tectonic plates. Beneath these plates is the mantle, where cooler liquid rock sinks and hot liquid rock rises, causing the tectonic plates to move. The plates can move side-by-side, towards or away from each other, depending on the movement of the mantle below. Most volcanoes, mountains and earthquakes are found along boundaries of tectonic plates.

THE ANATOMY OF A VOLCANO

WITH A MIGHTY ROAR, A RED-HOT FOUNTAIN OF LAVA EXPLODES
from the mountain. The lava flows, pools and cools into stunning forms, shaping the landscape for years to come. A volcano is an opening in the Earth's surface that enables hot molten rock, gas and ash to burst from the Earth's mantle. This is called an eruption.

SHIELD VOLCANO

STRATOVOLCANO

MALEO BIRD, INDONESIA

HOT HOT HOT!

Liquid or molten rock underneath the Earth's crust is called **magma**, but once it comes to the surface we call it **lava**. Lava can be as hot as 1,250°C!

While they can be split into various categories, there are two basic types of volcano: **shield volcanoes** and **stratovolcanoes**. Shield volcanoes are flatter, less explosive and have eruptions consisting mostly of runny **lava flows**.

Stratovolcanoes have steeper sides and larger, more explosive eruptions. As well as creating bigger **ash clouds** and **lava fountains**, they are capable of producing **pyroclastic flows**. These are very fast-moving clouds of extremely hot ash, lava fragments and gas. Pyroclastic flows are very dangerous and destructive.

The explosiveness of an eruption is determined by how easily natural gases are able to escape from the magma. The thicker and gooier the magma, the more gas bubbles become trapped over time. This causes pressure inside the magma chamber to build up until eventually… BANG!

Not all volcanoes are still active. **Extinct volcanoes** will never erupt again. **Dormant volcanoes** haven't erupted in a very long time but might erupt again in the future.

LIVING WITH VOLCANOES

You may think that with such explosive tendencies, volcanoes aren't very compatible with life. However, areas with volcanic activity can be excellent places for nature to thrive. The soil around volcanoes tends to be rich in **minerals** and incredibly fertile – perfect for plant life.

Some animals have adapted to live in volcanic environments too. Maleo birds sometimes bury their eggs in volcanic soil to keep them warm. When their chicks hatch, they dig their way out of the ground. What a place to call home!

PLATE KEY

CROSS-SECTION OF A VOLCANO

1. ASH CLOUD **2.** LAVA FOUNTAIN **3.** VOLCANIC BOMBS **4.** CRATER **5.** VENT **6.** SECONDARY VENT

7. CONDUIT **8.** BRANCH PIPE **9.** LAYERS OF ASH FROM PREVIOUS ERUPTIONS **10.** MAGMA CHAMBER

LAYERS OF HISTORY

WE CAN LEARN ALL ABOUT THE EARTH'S HISTORY FROM STUDYING THE GROUND BENEATH OUR FEET.

From major weather events to the activities of dinosaurs – layers of rock tell the story of our world.

IGNEOUS ROCK

METAMORPHIC ROCK

SEDIMENTARY ROCK

JURASSIC AMMONITE IN LIFE

HOW TO READ ROCKS

There are three basic categories of rock and each one tells the story of how it was made: igneous, metamorphic and sedimentary.

Igneous rocks are formed from lava or magma that has cooled and become solid, showing a history of tectonic activity in the area. They are very hard to break, and will often be dark in colour. Some rocks, such as pumice stones, have a holey texture, created as gas escapes from cooling lava.

Metamorphic rocks were originally other kinds of rock, but have been warped and transformed by heat and pressure deep underground. They will sometimes have interesting patterns from this transformation. Marble is an example of a metamorphic rock.

Sedimentary rocks are created from **sediment**, which is made of tiny bits of rock and **organic matter**, broken down through **erosion**. When sediment is compacted through pressure, sedimentary rock, such as limestone, is formed. Sedimentary rocks can often be identified by their layered appearance. Sometimes larger pieces of organic matter, like shells and bones, get trapped between layers, and are preserved as **fossils**.

BECOMING A FOSSIL

Fossils are found in layers of sedimentary rock. They are formed when sediment settles in and around a dead animal or plant – organic matter. This happens in places with slow-flowing water such as shallow seas, lakes and rivers.

Soft bits of the trapped organic matter decompose quite quickly. The hard bits that are left behind, such as bones and shells, are more likely to become fossilised because they take a long time to decompose.

If the conditions are right, the organic matter will be turned into stone, in a process called **petrification**. When the sediment becomes rock, minerals settle into all the tiny cracks and holes of the trapped organic matter. Over time, the original bits of organic matter dissolve, and the minerals in the water fill in the gaps, creating a beautiful rock copy – a fossil!

A JOURNEY THROUGH TIME

The layers of the Earth's crust can act as a timeline. Generally, the deeper the layer of rock, the older it is but this is not always the case. Layers are often warped, shifted and exposed by geological processes like earthquakes and erosion. This can give us easier access to study them.

The geological timeline is split into periods, each of which can stretch over millions of years. The end of each period marks a significant change to the Earth's climate and/or environment, which is usually the result of a big event that causes mass **extinction** and leads to a dramatic shift in the types of animals and plants that come after.

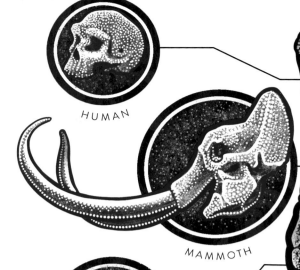

HUMAN

MAMMOTH

OREODONT

CRAB

VELOCIRAPTOR

CRINOIDS

ALLOSAURUS

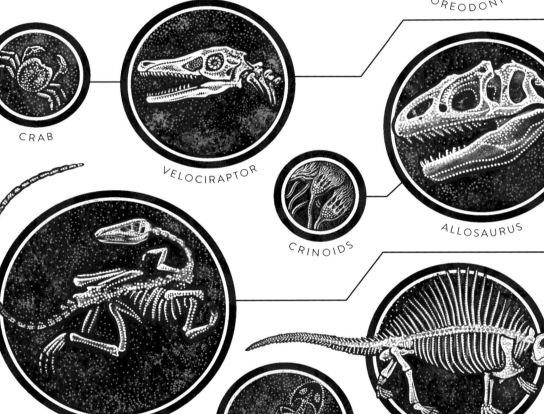

COELOPHYSIS

AMPHIBIAMUS LYELLI

DIMETRODON

FERN

EARLY CONIFER

STARFISH

BRACHIOPOD

NAUTILOID

TRILOBITE

QUATERNARY
Present - 2.6 million years ago

NEOGENE
2.6 - 23 million years ago

PALEOGENE
23 - 66 million years ago

CRETACEOUS
66 - 145 million years ago

JURASSIC
145 - 201 million years ago

TRIASSIC
201 - 252 million years ago

PERMIAN
252 - 299 million years ago

CARBONIFEROUS
299 - 359 million years ago

DEVONIAN
359 - 419 million years ago

SILURIAN
419 - 443 million years ago

ORDOVICIAN
443 - 485 million years ago

CAMBRIAN
485 - 539 million years ago

PRECAMBRIAN
539 million - 4.6 billion years ago

THE MAKING OF MOUNTAINS

IMPRESSIVE, EXCITING & FULL OF ADVENTURE,
mountains can be found all over the world.
But how are they made?

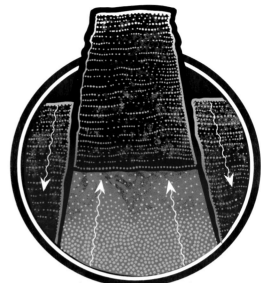

BLOCK MOUNTAINS

FOLD MOUNTAINS

COLLIDING PLATES

Fold mountains are the most common type of mountain. They are formed in the place where two tectonic plates push against each other, warping the Earth's crust into folds from the pressure. The Himalayan Mountains in Asia are an example of fold mountains.

CRACKED EARTH

Block mountains, also called fault-block mountains, are formed when chunks of the Earth's crust are forced up or down along **fault lines**. Fault lines are cracks in the Earth's surface or where tectonic plates move against each other. The Harz Mountains in Germany are an example of block mountains.

MOVEMENTS OF MAGMA

Dome mountains are formed from magma pushing the ground upwards, and then cooling to form igneous rock. **Volcanic mountains** are formed by volcanic eruptions. They can be active, dormant or extinct. Mount Fuji in Japan is an example of an active volcanic mountain.

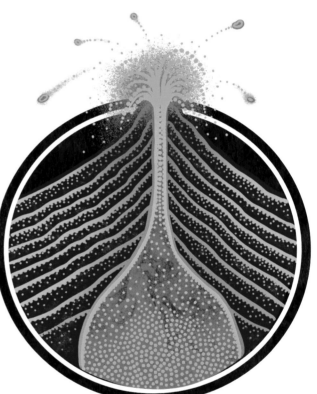

VOLCANIC MOUNTAINS

DOME MOUNTAINS

CARVED BY THE ELEMENTS

Mountains are also shaped by their environment. A plateau is a flat upland area and this can be 'dissected' by erosion of deep valleys to create the appearance of a mountainous landscape, sometimes referred to as **plateau mountains**. The Blue Mountains in Australia are an example of plateau mountains.

PLATEAU MOUNTAINS

PLATE KEY

1. MOUNT EVEREST The highest mountain in the world at 8,849 metres above sea level **2. TENZING NORGAY** & **3. EDMUND HILLARY** The first recorded people to reach the top of Mount Everest were Sherpa mountaineer Tenzing Norgay from Nepal and Edmund Hillary from New Zealand in 1953 **4. HIMALAYAN IBEX** An animal native to the Himalayas that is able to climb extremely steep cliffs due to its suction-cup-like hooves

EARTHQUAKES

YOU FEEL A SHARP JOLT. EVERYTHING SHAKES. YOU FEEL AS IF YOU ARE AT SEA, EVEN THOUGH you're on land. What's going on? Every 30 seconds, an earthquake happens somewhere on Earth. Most are very small and often go entirely unnoticed, but the bigger ones can cause a lot of damage and destruction.

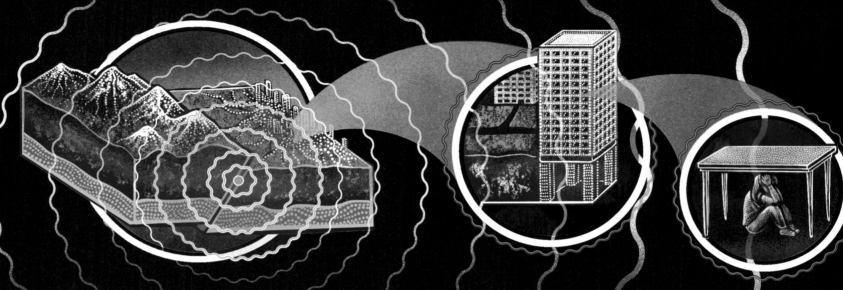

WHAT CAUSES AN EARTHQUAKE?

Tectonic plates are always on the move – albeit extremely slowly. However, they sometimes get stuck due to **friction**. Imagine the palms of your hands are tectonic plates. First, hold them tightly against each other, then try to slide them past each other – you might find that they stick, and the movement is jumpy. Tectonic plates also have this jumpy movement as they overcome the force of friction, causing an earthquake.

PREDICTING, PREPARING, SURVIVING

It's usually easy to predict where earthquakes will occur, as they happen close to the tectonic fault lines of the Earth. What is harder to predict is when they will happen, and how big they will be. For countries that experience earthquakes more often, there are ways to minimise the damage. Some buildings are designed with rubber foundations to absorb the shock, and reinforced with steel frames that can bend and sway with the motion of the ground, but these are expensive to build. Practice drills are held regularly to go through what to do in the case of an earthquake, such as hiding under a desk to protect yourself from falling debris.

GIANT WAVES

Many earthquakes happen out at sea. Extra-large earthquakes can displace a lot of water very suddenly, making a chain of particularly huge waves, called a **Tsunami**. These can be very destructive, as they cause mass flooding and the powerful force of the waves can cause serious damage when they collide with buildings.

Tsunamis move very quickly but they can sometimes be predicted a few hours before they hit a coastline if an earthquake is detected out at sea. Around 80% of Tsunamis originate from a particular place in the Pacific Ocean called the 'Ring of Fire'. The Pacific Tsunami Warning System uses sensors to track earthquakes that occur there, to predict when a Tsunami will hit, saving lives.

PLATE KEY

1. NORTH AMERICAN TECTONIC PLATE   **2.** EURASIAN TECTONIC PLATE   **3.** MID-ATLANTIC RIDGE

Rift valleys are huge splits in the Earth created when two tectonic plates are moving apart from each other. Earthquakes are common around these plate boundaries. Thingvellir Rift Valley in Iceland is the only place in the world where you can walk between two continental tectonic plates! On one side is the North American tectonic plate, and on the other is the Eurasian tectonic plate. They move incredibly slowly, at a rate of around one to two centimetres a year. Iceland experiences thousands of earthquakes a year due to the movement of the plates, however, these are generally just small tremors that are barely felt.

GEYSERS AND HOT SPRINGS

A CROWD OF PEOPLE CAUTIOUSLY SURROUNDS A MUDDY PUDDLE IN THE GROUND,
kept at a safe distance by a rope fence. There's a suspiciously eggy stench in the air.
They've been gazing at this puddle for a long time. All of a sudden, a hot fountain
of water and steam erupts. The crowd cheers with delight.

A PROPER GEYSER

Geysers are naturally occurring fountains of hot water found in areas of volcanic activity. They tend to erupt periodically, each geyser having a different length of interval between eruptions.

Old Faithful in Yellowstone National Park, USA, erupts once every 35–120 minutes, and its water jet can shoot up to 56 metres high!

TARDIGRADE

GEYSER

TRAPPED GAS

A geyser is supplied by a natural underground plumbing system. Water enters this system by soaking into the ground, where it collects in an underground tube running close to a magma chamber. The **groundwater** heats up and begins to boil, turning into gas, and building pressure inside chambers called bubble traps. Eventually the gas forces its way up the tube, pushing any water ahead of it and erupting above ground as a jet of hot water and steam – a geyser.

When you approach an area with geysers, you might find the air smells a bit like rotten eggs. Don't blame it on the dog though, this is due to high levels of a rather smelly mineral in the water called sulphur.

PRESSURE BUILDS FROM BOILING WATER

BUBBLE TRAP

MIGHTY MICRO-ANIMALS

While the boiling hot temperatures of geysers would kill most life forms, tiny animals called tardigrades have been found living in their waters. These microscopic animals are famous for surviving in extraordinarily inhospitable places. They are able to withstand extreme temperatures, pressures and starvation. They have even survived outer space!

GROUNDWATER RUNS CLOSE TO MAGMA, AND HEATS UP

MAGMA

A RELAXING BATH

All geysers are a kind of **hot spring**, but not all hot springs are geysers. Some springs present as pools of hot water. These occur in a similar way to geysers, where the groundwater heats up as it comes close to hot magma underground. The difference is that the water is able to heat up more slowly without the presence of bubble traps, and is therefore less explosive. People have been bathing in the pools of hot springs for all of human history.

MONKEY SEE, MONKEY DO!

In Jōshin'etsu-kōgen National Park, snow monkeys can be seen relaxing in a hot spring pool built just for them. It is said they saw humans soaking in a hot spring pool in the 1960s, and realised it was a great way to warm up from their cold, snowy environment. Generations of monkeys have been enjoying the warm waters ever since!

CAVES AND CRYSTALS

'ECHO!' 'ECHO!' 'ECHO!' YOUR VOICE BOUNCES OFF THE WALLS OF THE CAVE.

Spikes hang from the ceiling and rise from the ground. Crystals sparkle from the walls in the torchlight.

MAKING A CAVE

Most caves are formed when rainwater, which is slightly acidic, soaks into the ground and dissolves sections of soft rock, such as limestone. A bit like when you add sugar to a cup of tea and it seems to disappear, minerals from the rock combine with the water into a mixture and flow away. This process is known as **weathering**. Caves that form this way are called **solution caves**.

Some caves form in cliffs along the coast as crashing waves wear holes in the rock over time. These are called **sea caves**.

When lava flows out from a volcano, its outer surface begins to cool and crust over while hot lava continues to flow underneath. This leaves a tube-like cave behind, known as a **lava tube**.

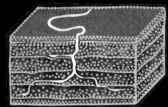

1. WATER SEEPS INTO THE GROUND

2. WATER DISSOLVES ROCK, FORMING A CAVE

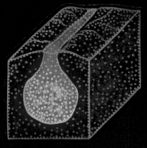

1. LAVA FLOWS WHILE SURFACE SOLIDIFIES

2. LAVA CONTINUES TO FLOW, LEAVING A LAVA TUBE

STALACTITE

STALAGMITE

COLUMN

A STAB IN THE DARK

In a solution cave, water carrying minerals slowly drips from the ceiling, leaving behind, or depositing, tiny bits of mineral with each drop. Rocks are made from minerals so, over a long time, interesting rock structures begin to build up. In a limestone cave, you might see spikes hanging from the ceiling like icicles, these are called **stalactites**. There might also be spikes sticking up from the floor, called **stalagmites**. These are created when drips hit the floor and deposit a little more of the minerals the water was carrying.

A good way to remember which one is which is to imagine that the 'g' in stalagmites stands for 'ground', and the 'c' in stalactites stands for 'ceiling'. Sometimes, they grow large enough to connect in the middle, forming a column.

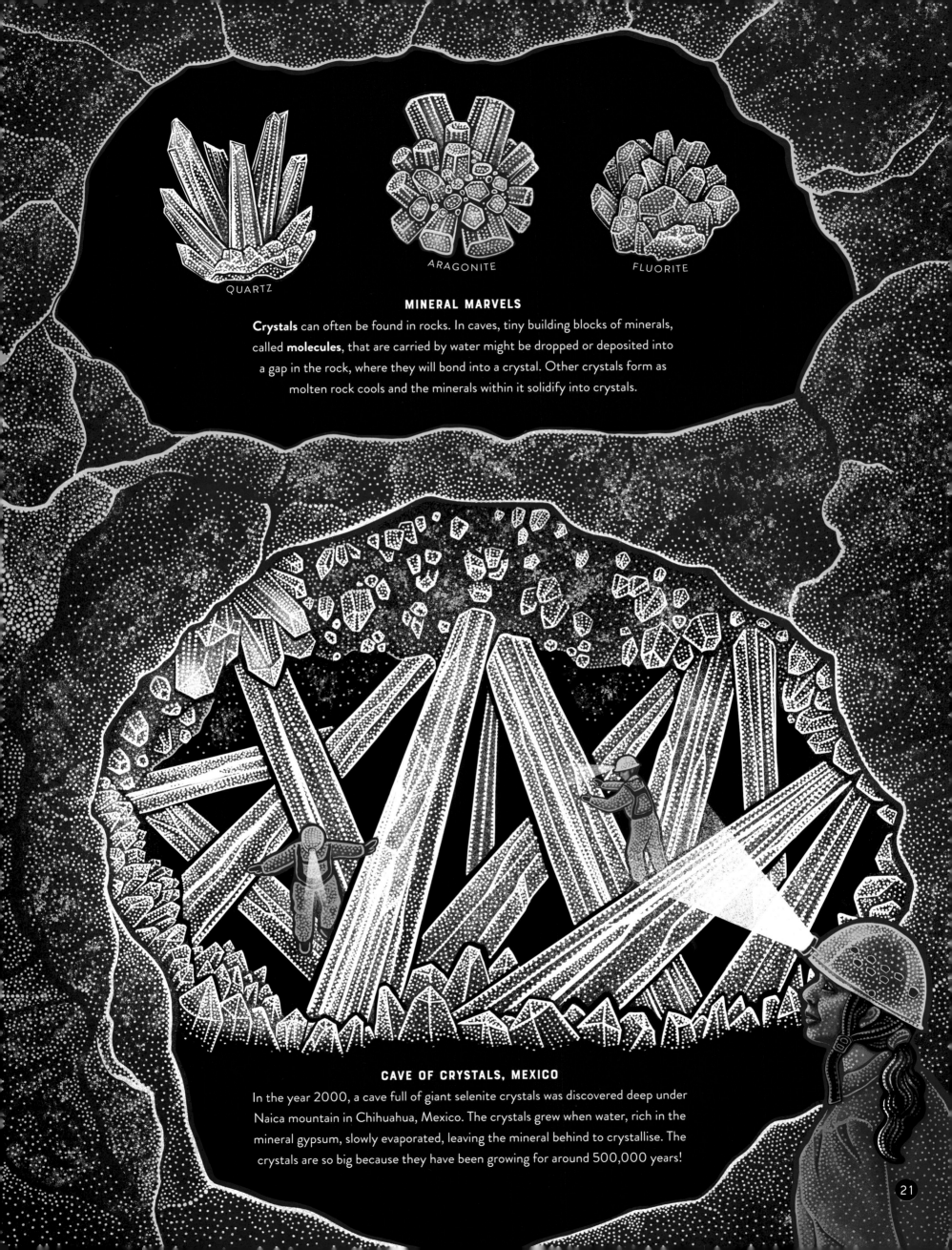

QUARTZ

ARAGONITE

FLUORITE

MINERAL MARVELS

Crystals can often be found in rocks. In caves, tiny building blocks of minerals, called **molecules**, that are carried by water might be dropped or deposited into a gap in the rock, where they will bond into a crystal. Other crystals form as molten rock cools and the minerals within it solidify into crystals.

CAVE OF CRYSTALS, MEXICO

In the year 2000, a cave full of giant selenite crystals was discovered deep under Naica mountain in Chihuahua, Mexico. The crystals grew when water, rich in the mineral gypsum, slowly evaporated, leaving the mineral behind to crystallise. The crystals are so big because they have been growing for around 500,000 years!

THE **POWER** OF **ICE**

ICE IS ONE OF THE MOST POWERFUL SHAPERS OF EARTH.

AN AGE OF ICE AND SNOW

In its 4.5 billion-year history, the Earth has gone through long periods of cooler temperatures, known as **ice ages**. Technically, we are still in an ice age, which began 3 million years ago!

What we might commonly think of as an ice age – freezing temperatures and large amounts of ice and snow – is called a **glacial period**. The last glacial period ended around 10,000 years ago, when life looked very different and mammoths still walked the Earth! As the name suggests, in these periods it gets so cold that **glaciers** are formed.

THE FRILLY FJORDS OF NORWAY

MARCH OF THE ICE GIANTS

Glaciers are enormous bodies of ice found in the coldest parts of the world. They are formed over hundreds of years as layer upon layer of snow becomes so thick that it turns into ice under its own weight.

The movement of a glacier shapes the landscape. It acts like an incredibly slow-moving river, eroding the land beneath it. Due to gravity, glaciers are always moving downhill, though only by a few centimetres a year. Their sheer size and weight mean they tear great channels or valleys out of the Earth's crust, not letting anything alter their downward course.

Glacial erosion sometimes appears on maps as 'frilly' edges of coastline. This can be seen in Norway, where glaciers have created fjords – long narrow inlets of water with steep mountains or cliffs on either side.

AMAZING EXPANDING ICE

Glaciers also erode the landscape through a process called **freeze-thaw weathering**.

When water freezes, it expands, taking up more space than it did as a liquid. Freeze-thaw weathering occurs when water freezes in the cracks of a rock and expands, creating a force powerful enough to split open even the largest of rocks.

Sometimes this phenomenon appears in rare and interesting ways. 'Frost flowers' can form when the water inside a plant's stem freezes and expands, forcing its way out in stunning petal-like formations of ice.

WATER GETS TRAPPED
IN CRACKS OF ROCK

WATER FREEZES
AND EXPANDS

ROCK BREAKS

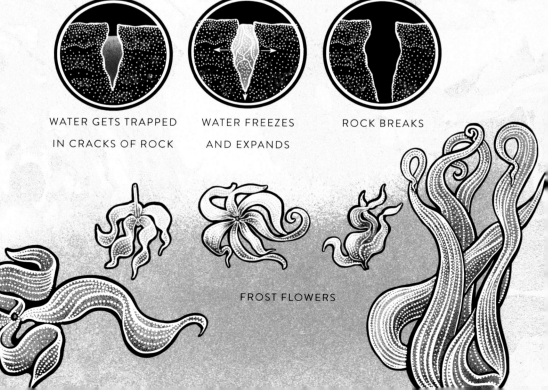

FROST FLOWERS

PLATE KEY

GLACIER BAY, ALASKA

1. MOUNTAIN VALLEY GLACIER **2.** ICEBERGS *Chunks of ice that break off from the glacier* **3.** HARBOUR SEAL

We are currently in an **interglacial period**, when glaciers recede or melt. However, due to **human-induced climate change**, the world's glaciers are melting at a faster rate than they should be. This is a big problem, as it will cause sea levels to rise and lead to more extreme weather events.

MARINE MARVELS

YOU ARE ON A BOAT AT SEA, LOOKING OUT AT THE VIEW AHEAD.

The sea around you is vast with no land in sight. You look down into the water, and it seems to be so deep, dark and mysterious. Did you know that water covers around 70% of the Earth's surface?

THE MOON AND THE TIDES

If you've ever been to the beach, you might have noticed that the edge of the sea comes in and out throughout the day, forcing you to move your towels and bags to prevent them from getting soggy! This is known as the tide cycle. The tide cycle is caused by the interaction between the pull of the moon's gravity, and Earth's own gravity. As a consequence, the ocean will bulge slightly on the area of Earth that is facing the moon, while on the opposite side of the Earth the ocean will bulge away from the moon. As the Earth rotates, the areas of water being affected change and so the water settles back down and it is this movement of the water that we see as tides.

WHIRLING WHIRLPOOLS

Sometimes, the movement of the tides can cause whirlpools, which are rotating bodies of water. They happen when two currents of water travelling in opposing directions meet each other, or an obstacle, causing the currents to swirl around in a spiral. A maelstrom is a big whirlpool. Most whirlpools aren't very strong, but some can form a downdraft, or a vortex. This means everything is sucked downwards, like bath water going down a plug hole. Maelstroms with a strong downdraft can be hazardous.

UNDERWATER VOLCANOES AND VENTS

Volcanoes aren't just found on land; eruptions happen on the seabed too. These are called **submarine volcanoes**. The lava from these eruptions can create new islands, such as those found in Hawaii, the Canary Islands and the Galapagos Islands. **Hydrothermal vents** tend to form around submarine volcanoes. They are created when seawater travels through cracks in the seabed and comes close to magma, which causes the water to heat up and rise back out of the seabed, heating the surrounding area. The vents often look like plumes of white or black smoke and are rich in minerals, which support living things in the dark depths of the oceans. It is possible that hydrothermal vents are where life first evolved on Earth!

PLATE KEY

CREATURES FOUND AT HYDROTHERMAL VENTS, NORTH WEST EIFUKU VOLCANO
1. GIANT TUBE WORMS 2. ZOARCID VENT FISH 3. POMPEII WORMS 4. HYDROTHERMAL MUSSELS
5. SHRIMP 6. VULCANOCTOPUS 7. GALATHEID CRABS

Hydrothermal vents are found so deep in the ocean that no sunlight can reach them, and their waters can heat up to 400°C!
While most living things wouldn't be able to survive these conditions, some creatures have adapted to thrive there.
Some cannot be found anywhere else on Earth except in the specific vent field where they live.

LIGHTS IN THE SKY

IT'S A CLEAR NIGHT IN NORWAY AND THERE ARE SWIRLING RIVERS OF GREEN LIGHT
in the dark sky above you. As you watch, you can see hints of purple join the dance too. It's a magical sight!

This phenomenon is called an aurora, or the 'polar lights' – as you are more likely to see them the closer you get to the North and South Poles.

These beautiful lights are caused by **solar wind**, a flow of **charged particles** coming from the Sun. But why do we only see them closer to the North and South Poles? It's because the Earth is a giant magnet.

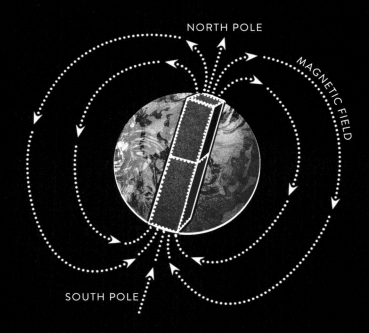

NORTH POLE

MAGNETIC FIELD

SOUTH POLE

A MAGNETIC DISPLAY
The Earth has a **magnetic field** thanks to its solid metal core. On an old-fashioned compass, the double-ended arrow aligns along the North-South axis when it is attracted to the magnetic North Pole.

The Earth's magnetic field acts like an umbrella, protecting our atmosphere from harmful space energy, like solar wind. But at the North and South Poles some of the solar wind is able to enter the atmosphere. The charged particles collide with gas particles in the atmosphere, which creates aurora light shows.

Particularly strong solar winds can sometimes interfere with technology, like the internet. Very occasionally a strong solar wind can cause a **geomagnetic storm**, making auroras visible all over the world!

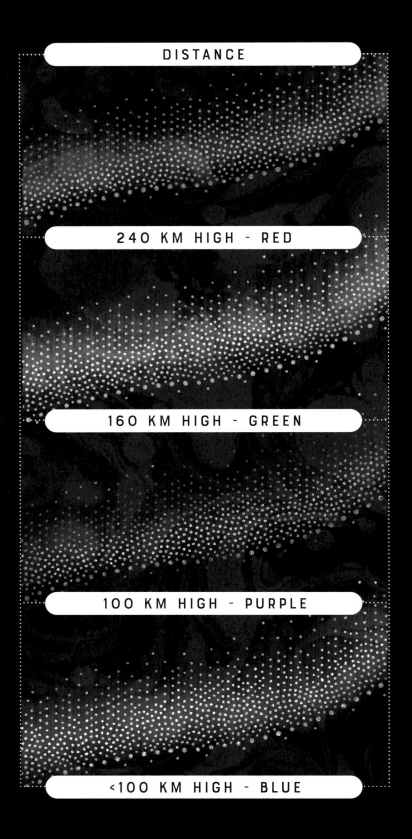

DISTANCE

240 KM HIGH - RED

160 KM HIGH - GREEN

100 KM HIGH - PURPLE

<100 KM HIGH - BLUE

ALL THE COLOURS OF THE SOLAR WIND
The most common colour of auroras is green, but solar wind colliding with various gases at different heights in the atmosphere creates other colours, including red, purple and even blue! The stronger the solar winds, the stronger the light show, with the possibility of a greater range of colours on display.

1. THE SUN **2.** SOLAR WIND **3.** EARTH'S MAGNETIC FIELD ACTS AS AN UMBRELLA AGAINST SOLAR WIND
4. THE POLES OF THE EARTH ARE WEAK POINTS OF ENTRY, ALLOWING SOLAR WIND TO INTERACT
WITH THE ATMOSPHERE, CAUSING THE POLAR LIGHTS **5.** EARTH

In this scene from Norway you can see the aurora borealis lighting up the sky.
Borealis refers to lights in Earth's northern latitudes. Those in the South are called aurora australis.

CLOUD SPOTTING

SOFT BILLOWING CLOUDS MOVE ACROSS THE SKY ON A WINDY DAY.

One looks like a dog, another looks like a dolphin! There are
lots of conditions that make clouds look different.

....

THE MAKING OF CLOUDS

Clouds are part of the **water cycle**. All water on Earth has been
in a continuous cycle, moving from land, to sky, to land again, for around
3.8 billion years. That means the water you drink today might have once
been peed out by a dinosaur! Luckily, the water cycle is a brilliant natural
purification system, making water as good as new.

1. EVAPORATION

Water turns into a gas called **water vapour**
when heated, in a process called evaporation.
On a sunny day, the water on the surface
of oceans and rivers begins to evaporate,
floating up into the sky as water vapour.

2. CONDENSATION

Water vapour cools down again as it rises, turning
into tiny water droplets. This process is called
condensation. The droplets are so small and light
that they float in the air. This is what clouds are
made of. In colder conditions, the droplets
freeze into tiny ice crystals, which
make the clouds look wispier.

THE WATER CYCLE

4. RUN OFF

Water from rain and snow
needs somewhere to go. Some
of it soaks into the earth and
becomes groundwater. When rain
is especially heavy, a lot of it runs
off the land into rivers and oceans.

3. PRECIPITATION

The more water condenses in
the clouds, the bigger the droplets get,
and the heavier they become, until they fall
back to Earth in a process called precipitation.
Depending on the temperature, precipitation
might take the form of rain, snow or hail.

NAMING THE CLOUDS

The names of most types of cloud come from Latin words describing their features:

Cirrus or *cirro* - high up and wispy

Stratus or *strato* - layered, flat and smooth

Cumulus or *cumulo* - puffy and fluffy

Alto - mid-level clouds

Nimbus or *Nimbo* - rain clouds

STORMY SKIES

A SUDDEN FLASH LIGHTS UP THE DARK, CLOUDY SKY.
A moment later you hear a loud rumble of thunder. Heavy rain
thrashes down and the wind howls. It's a thunderstorm!

....

A ZAP AND A CLAP

Thunderstorms happen when hot, moist air rises and cools
quickly as it meets the colder upper atmosphere. This causes
the water vapour to condense faster than usual, creating thick
storm clouds and heavy rain, and sometimes even icy hail.

Lightning is the release of electricity from the atmosphere
to the ground. During a storm, the water droplets and ice
crystals that form clouds move around and bump into
each other, creating **static electricity**. This static
charge builds up until it is eventually released
to the ground as **lightning**.

TOTALLY HAIR RAISING!

Have you ever rubbed a balloon against your hair? It creates static
electricity, which causes your hair to stick up. If you're ever out in rainy
weather and your hair starts to stick up, that's a sign that lightning
is about to strike. You should head back indoors!

Thunder is the sound made as lightning super-heats the air to five
times the temperature of the Sun! The shorter the pause between
seeing lightning and hearing thunder, the closer the lightning is.
This is because light travels faster than sound.

HOW A STORM SYSTEM FORMS

A HURRICANE FROM ABOVE

SWIRLY STORM SYSTEMS

You may have heard the expression 'the eye of the storm', the term for the strangely calm environment that can be found at the very centre of a tropical storm such as a hurricane. Tropical storms are created when warm, damp air rises up off tropical seas, which creates an area of low pressure. As the air then cools and water vapour condenses into clouds, it is pushed sideways by the warm, damp air rising below it, creating a rotating system of strong winds and rain.

WHEN DOES A STORM BECOME A HURRICANE?

Since storms are caused by evaporation, the hotter and more humid an area, the more severe they tend to be. This is why large thunderstorms are more common in tropical areas. We categorise storms by how strong the winds are. When wind speeds reach 74mph or more, a storm will be labelled as a **hurricane** if it started over the North Atlantic, central North Pacific or Northeast Pacific Ocean, **cyclone** if it started above the Indian or South Pacific Ocean and **typhoon** if the hot, moist air was drawn up off the Northwest Pacific Ocean.

TWISTY TORNADOES

Tornadoes, often called twisters, are spinning, funnel-like columns of air that reach from a storm cloud to the Earth's surface. They are usually short-lived but, as they can rotate at up to 300mph, they can cause a lot of destruction in that short amount of time!

TRICKS OF THE LIGHT

DID YOU KNOW THAT RAINBOWS ARE AN OPTICAL ILLUSION?

An optical illusion is something that you see, but is not really there.

....

OPTICAL ILLUSIONS

We see rainbows when light shines through water droplets at a certain angle. This is why you can sometimes see rainbows in the mist of waterfalls, or even in the water from your garden hose in the summer! Although it appears white, light from the Sun is actually a mixture of every different colour. When it hits a raindrop, for example, it splits into a rainbow of colour. This is due to a process called **refraction**.

BENDING LIGHT

Refraction is when light changes direction and so appears to bend. This happens when it passes between different transparent materials, such as air, water and ice. Light travels at different speeds through each material, changing the angle of its path. You can test this for yourself by putting a straight object, like a pencil, into a glass of water. It will look like the straight object is bent!

MIRAGE

Sometimes people see pools of water in a desert that aren't there. This phenomenon is known as a mirage. You might see one for yourself as the air shimmers above a hot tarmac road. A mirage can occur when the air next to the ground is very hot, and there is cooler air just above it. The difference in temperature means that the layers of air have different densities. As the light travels through each layer, it alters course – or bends – so much that it acts like a mirror reflecting the sky above. However, our eyes trace the reflected light back in a straight line, so the image of the blue sky is displaced onto the ground, making it look like shimmering water!

SO ATMOSPHERIC!

Sometimes ice crystals suspended in the sky can cause light to reflect and refract in beautiful ways. Light pillars can extend into the sky from light sources. Halos of light can shine around the Sun, as well as create 'sun dogs' – coloured lights on either side of the halo.

LIGHT PILLARS

SUN DOGS

'FIREFALL' HORSETAIL FALL, YOSEMITE, USA

During the month of February, at sunset, a waterfall called Horsetail Fall looks like fiery lava cascading from the top of a cliff edge. This is an optical illusion created by the golden light of the setting Sun hitting the water at the perfect angle!

ADVENTURES IN GEOSCIENCE

FROM STANDING AT THE EDGE OF LAVA LAKES,
to cataloguing precious fossils in museums, we have geoscientists
to thank for all we know about Earth's phenomena.

·······································

VOLCANOES

Getting up close to an active volcano might not appeal to everyone,
but for these scientists, it's their life's work! A volcanologist
studies volcanoes and their eruptions. They visit
volcanoes to monitor their activity, and collect
samples of rock and lava to study.

EARTHQUAKES

Seismologists study the structure of the Earth in order to
understand and try to predict earthquakes. They monitor
seismic changes using specialised equipment and map fault
lines in the Earth's surface. They also develop early warning
systems for places that are prone to earthquakes.

OCEAN

Oceanographers study the ocean in all its complexity.
They study the geology of the ocean floor, its ecosystems,
the chemicals in the water, ocean currents and how all of these
things affect each other. While a lot of their work happens in
the laboratory, they will frequently make research trips out
to sea on boats, scuba dive to visit coral reefs, and descend
into the depths in submersibles.

WEATHER

Ever fancied chasing a storm? How about a tornado?
Meteorologists study short-term weather changes; you may have seen them on the news, though not all meteorologists are on TV! With their knowledge of weather patterns and special equipment, they are able to predict with a certain amount of accuracy what the weather is going to be like in the immediate future.

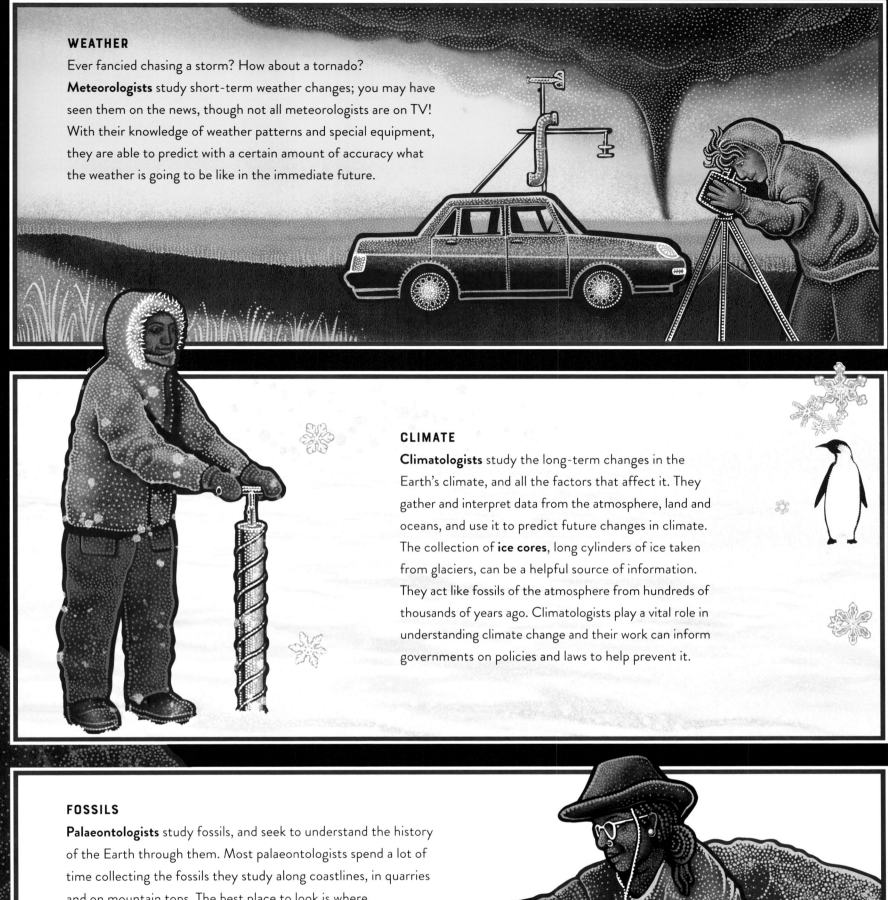

CLIMATE

Climatologists study the long-term changes in the Earth's climate, and all the factors that affect it. They gather and interpret data from the atmosphere, land and oceans, and use it to predict future changes in climate. The collection of **ice cores**, long cylinders of ice taken from glaciers, can be a helpful source of information. They act like fossils of the atmosphere from hundreds of thousands of years ago. Climatologists play a vital role in understanding climate change and their work can inform governments on policies and laws to help prevent it.

FOSSILS

Palaeontologists study fossils, and seek to understand the history of the Earth through them. Most palaeontologists spend a lot of time collecting the fossils they study along coastlines, in quarries and on mountain tops. The best place to look is where a rock face is being eroded away!

LEGENDS OF THE EARTH

EVER LOOKED AT A STRANGELY SHAPED HILL, AND MADE UP A STORY OF HOW IT CAME TO BE?

Before humans turned to science for answers, many cultures sought to understand
the world around them through legends and stories.

WHAT CAN THESE STORIES TELL US?

Ancient legends sometimes reference real geological events from history, and they can contain
important clues for **geologists** and **archaeologists** that lead to certain locations or areas of particular
scientific interest. Far from being simple tales of fantasy, these stories can hold deep cultural and
spiritual meaning for the people that share them, and connect them with the land they call home.

THE ANCIENT GREEK GODS

The ancient Greeks believed that many natural events
and phenomena in the world around them were created
by the activities of the gods.

Earthquakes and tremors were caused by struggling
giants who were trapped underground by the god Zeus,
while volcanoes were said to be chimneys belonging to
the forge of the blacksmith god Hephaestus. Volcanic
eruptions sometimes make loud metallic noises, which
can sound like the clanging of a blacksmith's hammer.

THE RACE OF THE HAWAIIAN GODDESSES

In Hawaiian folklore, Madame Pele is a fire goddess and the creator of
the islands. She is known for her fiery temper, and volcanic eruptions are
said to be caused by her anger and earthquakes happen when she stomps
her feet. She is both a destructive and creative goddess, as her eruptions
create new islands and fertile land.

Many stories featuring Pele point to real geographic phenomena. For
example, the dense basalt rock found on the mountain Mauna Kea is said to
have been created in a battle between Madame Pele and the snow goddess
Poliʻahu. After Pele lost a hōlua sledding competition against Poliʻahu, she
became upset and started throwing lava. Poliʻahu fought back by cooling
the lava flows with her snow. This is interesting, as basalt is indeed formed
by the rapid cooling of lava.

A HILL CARVED BY A GIANT BEAR

Bear Lodge Butte, also known as Devil's Tower, is a steep hill in Wyoming, USA, made of rock shaped through volcanic activity. It's known for its strangely stripy formation, as if it had been scraped by enormous claws.

At least 24 Native American tribes have similar legends about how this hill came to be, though the exact details and endings differ. Most stories tell of the Great Spirit raising the hill out of the ground to save people who were being chased by a giant bear. The bear carved the stripes into the rock in its frustrated scramble to try to reach the people at the top.

THE NORSE LEGEND OF FIMBULWINTER

There is a Norse legend of an 'unending' winter, called 'Fimbulvetr' or 'Fimbulwinter'. Researchers think it might originate from Viking memory of a real enormous volcanic eruption that took place in South America and sent so much ash into the atmosphere that it blocked out the Sun! This led to a winter in Scandinavia that lasted at least three years, affecting the growth of crops and many livelihoods. It would have felt like the winter was unending at the time!

A 37,000-YEAR-OLD STORY

Indigenous Australians have a rich tradition of oral storytelling. One story from the aboriginal Gunditjmara people is thought to describe an eruption that happened around 37,000 years ago. It tells of four giants who arrived in Southeast Australia. While three continued their travels, one stayed near the coast and turned into the volcano Budj Bim. His teeth became the lava, which dramatically changed the landscape.

No other eruptions have taken place in the area that could have inspired this story, and there is archaeological evidence that the Gunditjmara people may have been settled there for 50,000 years. It is thought to be the oldest story that continues to be passed from generation to generation through speech.

GLOSSARY

archaeologists People who study human history by searching for and examining human remains and historic artifacts.

ash cloud A cloud consisting of fragments of rock, minerals and volcanic glass that is produced when a volcano erupts.

atmospheric phenomena Phenomena that occur within Earth's atmosphere, including weather events like tornadoes and lightning, and optical phenomena such as aurora and rainbows.

aurora A natural display of coloured lights that sometimes appears in the night sky around the North and South Poles.

block mountains Mountains that form when a block of rock is forced upwards or downwards between faults – or cracks – in the Earth's crust.

charged particles Particles (extremely tiny pieces of matter) that have an electrical charge.

climate The average weather conditions in a particular location over a long period of time.

crystals A solid whose molecules are arranged in a repeating pattern. Many crystals are found in rocks and can have beautiful and unusual shapes.

cyclone A tropical storm that forms over the South Pacific or Indian Ocean.

dome mountains Mountains that form when magma pushes the ground upwards and then cools to form igneous rock.

dormant volcano A volcano that has not erupted in a long time but may erupt in the future.

erosion When rock or soil is worn away and transported by natural forces, such as wind or water.

extinct volcano A volcano that will never erupt again.

extinction When a species of animal or plant completely dies out.

fault lines Long cracks in the surface of the Earth.

fold mountains Mountains that form when tectonic plates are pushed together.

fossils Preserved remains or traces of animals or plants that lived long ago. They are often found buried in rock.

freeze-thaw weathering A type of erosion where water freezes and expands inside cracks in rock, causing the rock to break apart.

friction A force created between two surfaces when one object moves or rubs against another.

geologists People who study the structure of the Earth and the things it is made of.

geology The science of the structure of the Earth and the things it is made of.

geomagnetic storm A disturbance of the Earth's upper atmosphere caused by solar winds.

geoscientists People who study the science of the Earth.

geysers Hot springs that periodically shoot out jets of hot water and steam.

glacial period A long period of particularly cold temperatures that occurs during an ice age, in which large areas of the Earth are covered in glaciers.

glaciers Large, slow-moving rivers of ice.

groundwater Water found beneath the Earth's surface.

hot spring A spring of water that is naturally heated by underground volcanic activity.

human-induced climate change The Earth's temperature is rising faster than it should due to human activity, such as burning fossil fuels.

hurricane A tropical storm that forms over the North Atlantic, central North Pacific or Northeast Pacific Ocean.

hydrothermal vents An opening in the seabed that heated water flows from.

ice ages Long periods of colder global temperatures, with an increase of ice sheets and glaciers. An ice age can last millions of years.

igneous rocks Rocks that form when lava or magma cools and becomes solid.

interglacial period A stretch of time between glacial periods when global temperatures rise and glaciers melt.

lava Liquid or molten rock that flows onto the Earth's surface from a volcano or crack in the Earth's crust.

lava flows Streams of lava that pour out from a volcano during an eruption.

lava fountain A jet of lava that sprays out from a volcano during an eruption.

lightning A giant electrical spark caused by static electricity within a thundercloud, which is released to the ground.

magma Liquid or molten rock that flows below the Earth's crust.

magnetic field The area around a magnet where there is magnetic force.

metamorphic rocks Rocks that have been transformed from one type to another by heat and pressure.

minerals Solid natural substances found in the earth, such as metals or salt.

molecules The tiny building blocks that all materials are made of.

organic matter The remains of things that used to be alive.

petrification The process of organic matter being turned into a stony substance over time.

plateau mountains Flat upland areas that have eroded in a way that makes them resemble mountains.

purification The removal of contaminants to make something – such as water – clean and safe.

pyroclastic flows A dense, fast-moving mixture of hot rock fragments, volcanic ash, and gas that flows from volcanoes during certain eruptions.

refraction When light changes direction and appears to bend as it travels through one material to another.

rift valleys Large valleys or trenches that form between two tectonic plates as they move away from each other.

sea caves Caves formed in cliffs by sea waves eroding the rock over time.

sediment Small pieces of materials such as rock, minerals and the remains of dead plants and animals, which are transported by water or wind before settling in layers on the ground or seabed.

sedimentary rocks Rocks that form as layers of sediment build up and compress together over millions of years.

shield volcanoes Volcanoes with gently sloping sides and a domed shape. Their eruptions tend to be less explosive than those of stratovolcanoes.

solar wind A stream of charged particles that comes from the Sun and travels throughout the solar system.

solution caves Caves formed by acidic water slowly dissolving rock over time.

stalactites Spikes of rock that hang from the ceiling of caves, formed by the dripping of water containing minerals.

stalagmites Spikes of rock that stick out from the floor of caves, formed by the dripping of water containing minerals.

static electricity A type of electricity that builds up on the surface of a material or between materials, often caused by friction.

stratovolcanoes A cone-shaped volcano with steep sides. Their eruptions can be very explosive.

submarine volcanoes Volcanoes that erupt under the surface of the ocean.

tectonic plates Sections of the Earth's crust that fit together to form the surface of the Earth.

tsunami A giant ocean wave, usually caused by an earthquake or volcanic eruption under the sea.

typhoon A tropical storm that forms over the Northwest Pacific Ocean.

volcanic mountains Mountains formed by volcanic eruptions.

water cycle The continuous cycle that all water travels as it moves around the Earth in different states.

water vapour Water in its gas state.

weathering The wearing away or dissolving of rocks, soil and minerals.

INDEX

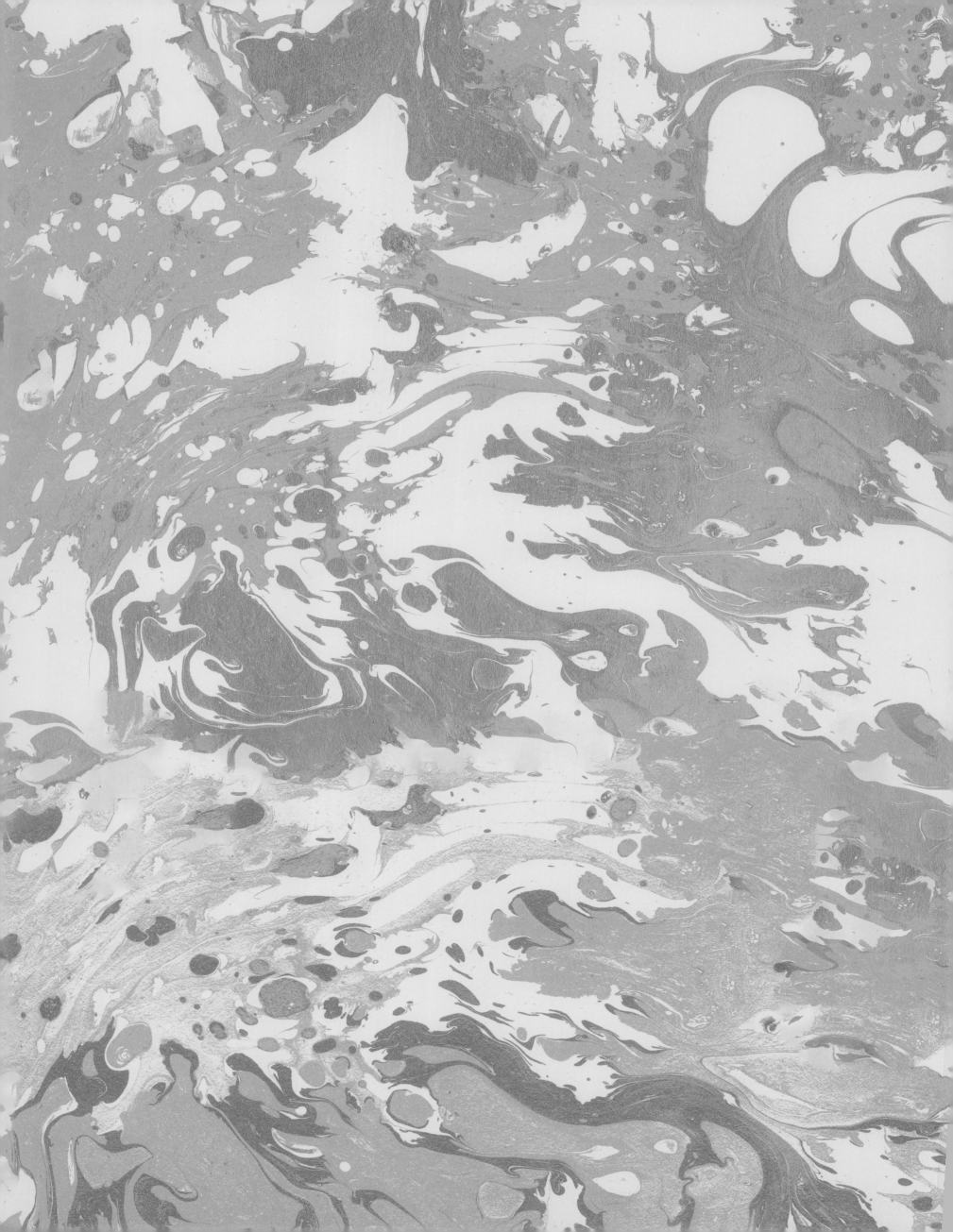

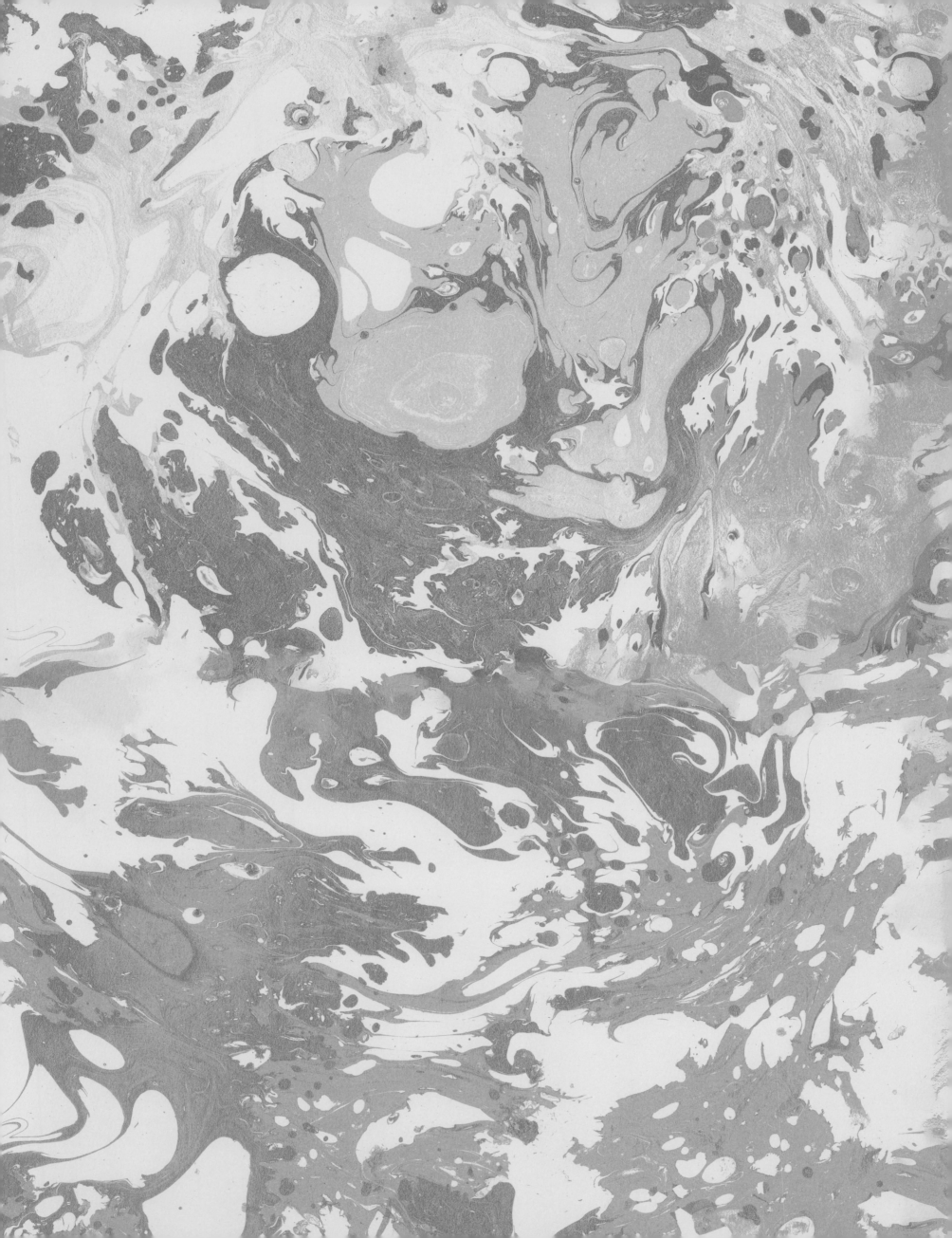